YOUR KNOWLEDGE HAS VALUE

- We will publish your bachelor's and master's thesis, essays and papers

- Your own eBook and book - sold worldwide in all relevant shops

- Earn money with each sale

Upload your text at www.GRIN.com
and publish for free

Bibliographic information published by the German National Library:

The German National Library lists this publication in the National Bibliography; detailed bibliographic data are available on the Internet at http://dnb.dnb.de .

This book is copyright material and must not be copied, reproduced, transferred, distributed, leased, licensed or publicly performed or used in any way except as specifically permitted in writing by the publishers, as allowed under the terms and conditions under which it was purchased or as strictly permitted by applicable copyright law. Any unauthorized distribution or use of this text may be a direct infringement of the author s and publisher s rights and those responsible may be liable in law accordingly.

Imprint:

Copyright © 2017 GRIN Verlag, Open Publishing GmbH
Print and binding: Books on Demand GmbH, Norderstedt Germany
ISBN: 9783668518971

This book at GRIN:

http://www.grin.com/en/e-book/372471/paradoxes-of-imaginary-unit-iota

Dharmendra Kumar Yadav

Paradoxes of Imaginary Unit Iota

GRIN Publishing

GRIN - Your knowledge has value

Since its foundation in 1998, GRIN has specialized in publishing academic texts by students, college teachers and other academics as e-book and printed book. The website www.grin.com is an ideal platform for presenting term papers, final papers, scientific essays, dissertations and specialist books.

Visit us on the internet:

http://www.grin.com/

http://www.facebook.com/grincom

http://www.twitter.com/grin_com

PARADOXES OF IMAGINARY UNIT IOTA

Dharmendra Kumar Yadav

Assistant Professor, Department of Mathematics

Shivaji College, University of Delhi, Raja Garden, Delhi

Abstract

In the paper some paradoxes of imaginary unit *iota* have been propounded. The properties found in the paper are more advanced than the previous ones. It has been proved that the imaginary numbers cannot be imagined within the set of real numbers. The representation of imaginary numbers (if possible) lie beyond the set of real numbers on real number line, which supports the concept of imaginary and circular number lines. Before concluding the paper one open problem has been presented for further research.

Key Words: Philosophy of Mathematics, Infinity, Imaginary Unit

MSC2000: 00A30, 03A05, 97F50

Introduction

Swiss Mathematician *Leonhard Euler* (1707-1783) introduced the imaginary unit 'iota' with symbol 'i' for the square root of (-1) with the property $i^2 = -1$ or $\sqrt{-1} = i$ in 1748. The existence of complex numbers were not completely accepted until the geometrical interpretation had been described by *Casper Wessel* in 1797 and *C. F. Gauss* in 1799 as points in a plane. Although the idea of representing a complex number by a point in a plane had been suggested by several mathematicians, it was *Argand's* proposal that was accepted. Gauss used it and proved "the Fundamental Theorem of Algebra" in his Ph.D. thesis in 1799 which had been given by *Albert Girard*. In 1833 Irish mathematician *Hamilton* introduced the complex number notation a + i b and made the connection with the point (a, b) in the plane although many mathematicians argued that they had found this earlier. In 1804 *Abbe Buee* came upon the same idea which *Wallis* had suggested that $\pm\sqrt{-1}$ should represent a unit line, and its negative, perpendicular to the real axis. Buee's paper was not published until 1806, in which year *Jean-Robert Argand* also issued a pamphlet on the same subject. It is to Argand's essay that the scientific foundation for the graphic representation of complex numbers is now generally referred.

A Look at Latest Works

Yadav [2008] proved that $i < 0$, $-i > 0$, $i < -\infty$ and $-i > \infty$. But he didn't discuss the signs of real number x and its effect, which lacked some properties. Based on these he introduced *imaginary number line* to represent imaginary numbers on it and then extended it to circular number line, which naturally wants some attention due to some undiscussed cases. On the other hand *Mabkhout* [2016] proved that "imaginary number merges to infinity and $i = -\infty$ as well as 'infinity' and 'imaginary' are equivalent".

Analysis of Imaginary Unit

Euler introduced the imaginary unit 'i' with the property $\sqrt{-1} = i$ and $i^2 = -1$. Since we have

$$\sqrt{-1} = \sqrt{-1 \pm i0} = \sqrt{\cos \pi \pm i \sin \pi} = \cos\frac{\pi}{2} \pm i\sin\frac{\pi}{2} = 0 \pm i = \pm i$$

$$\Rightarrow \sqrt{-1} = \pm i$$

and

$$\sqrt{-1} = \sqrt{-1 \pm i0} = \sqrt{\cos \pi \pm i \sin \pi} = \cos\left(\frac{2n\pi + \pi}{2}\right) \pm i\sin\left(\frac{2n\pi + \pi}{2}\right) = 0 \pm i;\ n = 0,1$$

From above we find that $\sqrt{-1} = \pm i$.

But it is not known that whether i > 0 or i < 0, therefore it shouldn't be taken as '+ i', ignoring the second one absolutely.

Discussion on Imaginary Unit

Although it is not known that a real number and an imaginary unit (or a complex number) can be compared or not, as two real numbers are compared with respect to law of trichotomy, we are proceeding in this direction using following properties:

Theorem 1: For non-zero real number x, the imaginary unit

(i) $i < \dfrac{x^2-1}{2x}$; for $x > 0$, and

(ii) $i > \dfrac{x^2-1}{2x}$; for $x < 0$.

Proof: Since x is a non-zero real number, let us consider the following cases:

Case-I: Let $i > x$. Then $i - x > 0$

$\Rightarrow (i - x)^2 > 0 \Rightarrow i^2 + x^2 - 2ix > 0 \Rightarrow 2ix < x^2 - 1$

$\Rightarrow i < \dfrac{x^2-1}{2x}$; for $x > 0$ and $i > \dfrac{x^2-1}{2x}$; for $x < 0$.

Case-II: Let $i < x \Rightarrow x - i > 0 \Rightarrow (x - i)^2 > 0 \Rightarrow x^2 + i^2 - 2ix > 0 \Rightarrow x^2 - 1 > 2ix$

$\Rightarrow i < \dfrac{x^2-1}{2x}$; for $x > 0$ and $i > \dfrac{x^2-1}{2x}$; for $x < 0$.

Thus we get the proof of the statement of theorem.

Theorem 2: For non-zero real number x, the imaginary unit

(i) $i > \dfrac{-(x^2-1)}{2x}$; for $x > 0$, and

(ii) $i < \dfrac{-(x^2-1)}{2x}$; for $x < 0$.

Proof: Since x is a non-zero real number, let us consider the following cases:

Case-I: Let $-i > x$. Then $i + x < 0$

$\Rightarrow (i + x)^2 > 0 \Rightarrow i^2 + x^2 + 2ix > 0 \Rightarrow 2ix > 1 - x^2$

$\Rightarrow i > \dfrac{(1-x^2)}{2x}$; for $x > 0$ and $i < \dfrac{(1-x^2)}{2x}$; for $x < 0$

$$\Rightarrow i > \frac{-(x^2-1)}{2x}; \text{ for } x > 0 \text{ and } i < \frac{-(x^2-1)}{2x}; \text{ for } x < 0.$$

Case-II: Let $-i < x \Rightarrow x + i > 0 \Rightarrow (x+i)^2 > 0 \Rightarrow x^2 + i^2 + 2ix > 0 \Rightarrow 1 - x^2 < 2ix$

$$\Rightarrow i > \frac{(1-x^2)}{2x}; \text{ for } x > 0 \text{ and } i < \frac{(1-x^2)}{2x}; \text{ for } x < 0$$

$$\Rightarrow i > \frac{-(x^2-1)}{2x}; \text{ for } x > 0 \text{ and } i < \frac{-(x^2-1)}{2x}; \text{ for } x < 0.$$

Thus we get the proof of the statement of theorem.

From the above two theorems, we conclude the following paradox:

First Paradox of Imaginary Unit

For a non-zero real number x, we have

(i) $i < \frac{(x^2-1)}{2x}$ and $i > -\frac{(x^2-1)}{2x}$ for $x > 0$, and

(ii) $i > \frac{(x^2-1)}{2x}$ and $i < -\frac{(x^2-1)}{2x}$ for $x < 0$.

Two Properties of Real Numbers

To compare real and imaginary numbers, we need some properties between a non-zero real number x and a real number of the particular form, as has been discussed below:

Theorem 3: For every non-zero real number $x > 0$, $x > \frac{x^2-1}{2x}$.

Proof: Since $1 > -x^2 \Rightarrow 1 - x^2 > -2x^2$ [Adding $-x^2$]

$$\Rightarrow \frac{1-x^2}{2x} > -x \text{ [Since } x > 0] \Rightarrow \frac{x^2-1}{2x} < x \text{ [Multiplying by -1]}$$

$$\Rightarrow x > \frac{x^2-1}{2x}.$$

Theorem 4: For every non-zero real number $x < 0$, $\frac{1-x^2}{2x} < -x$.

Proof: Since $1 > -x^2 \Rightarrow 1 - x^2 > -2x^2 \Rightarrow \frac{1-x^2}{2x} < -x$ [Since $x < 0$].

Relation Between Real Numbers and Imaginary Unit

Using the statement of four theorems, we get the following properties:

Theorem 5: The imaginary unit 'i' is greater than every negative real number and is less than every positive real number.

Proof: Let us consider two different cases:

Case-I: We have proved that for $x > 0$,

$$i < -\left(\frac{1-x^2}{2x}\right) \qquad (1)$$

$$i > \left(\frac{1-x^2}{2x}\right) \qquad (2)$$

and

$$\left(\frac{1-x^2}{2x}\right) > -x \qquad (3)$$

From (2) and (3), we have

$$i > -x \qquad (4)$$

From (1) and (3), we have

$$-i>\left(\frac{1-x^2}{2x}\right)>-x \Rightarrow -i>-x \Rightarrow i<x \qquad (5)$$

Case-II: Also we have proved that for $x < 0$

$$i>\left(\frac{x^2-1}{2x}\right) \qquad (6)$$

$$i<-\left(\frac{x^2-1}{2x}\right) \qquad (7)$$

and

$$\left(\frac{1-x^2}{2x}\right)<-x \qquad (8)$$

From (6) and (7) we have

$$i>\left(\frac{x^2-1}{2x}\right)>x \Rightarrow i>x$$

Since $x < 0$, putting $x = -y$ for $y > 0$, we get

$$i>-y \qquad (9)$$

From (7) and (8) we have

$$-i>\left(\frac{x^2-1}{2x}\right)>x \Rightarrow -i>x$$

Since $x < 0$, putting $x = -y$ for $y > 0$, we get

$$i<y \qquad (10)$$

From (4), (5), (9) and (10), we find that the imaginary unit 'i' is greater than every negative real number and is less than every positive real number.

Note: Since $i > -x$ and $i < x$. Multiplying by -1, we get $-i < x$ and $i < x$. Thus $\pm i < x$. Similarly multiplying by -1, we get $i > -x$ and $-i > -x$. Thus $\pm i > -x$. We can show it on the real number line as follows:

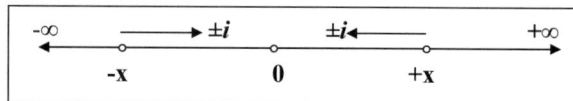

From above we conclude the following paradox:

Second Paradox of Imaginary Unit

The imaginary unit '±i' is both greater than as well as less than every non-zero real number.

Imaginary Unit as Origin

From the statement "the imaginary unit 'i' is greater than every negative real number and is less than every positive real number" indicate that 'i' must be zero if we limit ourselves in the real number system.

Whereas the statement "the imaginary unit '±i' is both greater than as well as less than every real number" indicates that the imaginary numbers don't lie on the real number line and they form another line possibly the imaginary number line and circular number line as were introduced by **Yadav** [2008, 2009].

Imaginary number line is composed of only imaginary numbers where as circular number line consists of both real number line and imaginary number line. On these lines, ±i behave like origin and its possible position is just opposite to the real number zero as shown below:

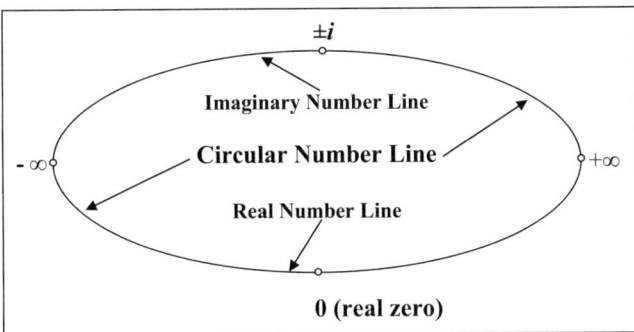

We have proved that i > - x and i < x for all non-zero real x. Taking $x \to 0$, we find that i > 0 and i < 0. We have also proved that ± i > - x and ± i < x for all non-zero real x. Taking $x \to 0$, we find that ± i > 0 and ± i < 0. These also strengthen the concept of imaginary origin (±i) on imaginary number line.

Limiting Value of Imaginary Unit Near Zero

Using above properties, let us consider the limiting values of iota near zero. We know that

(i) For x > 0, $i < \left(\dfrac{x^2 - 1}{2x} \right)$. Then for $x \to 0+$ we get $i < -\infty$.

(ii) For x > 0, we have $i > -\left(\dfrac{x^2-1}{2x}\right)$. Then for $x \to 0+$, we get $i > \infty$.

(iii) For x < 0, $i > \left(\dfrac{x^2-1}{2x}\right)$. Then for $x \to 0-$ we get $i > \infty$.

(iv) For x < 0, we have $i < -\left(\dfrac{x^2-1}{2x}\right)$. Thus for $x \to 0-$, we get $i < -\infty$.

Thus we find that the imaginary unit iota is greater than infinity (+∞) and is less than minus infinity (-∞).

Note: Multiplying both by (-1), we find that $-i > \infty$ and $-i < -\infty$. Thus
$$\pm i > \infty \text{ and } \pm i < -\infty.$$
Form this we conclude the following paradoxes.

Third Paradox of Imaginary Unit

The imaginary unit iota is both positive as well as negative.

Fourth Paradox of Imaginary Unit

The imaginary unit '±i' lie beyond the real numbers.

Note: The third and fourth paradoxes strengthen the possibility that '± i' behaves like origin (middle point) on imaginary number line, as zero behaves as a origin (middle point) on real number line.

Limiting Value of Imaginary Unit Near Infinity

Similarly as above, we know that

(i) For $x > 0$, $i < \left(\dfrac{x^2-1}{2x}\right)$. Then for $x \to \infty+$, we have

$$\lim_{x \to \infty+} \dfrac{x^2-1}{2x} = \lim_{x \to \infty+} \dfrac{1-\dfrac{1}{x^2}}{\dfrac{2}{x}} = \dfrac{1-0}{0} = \infty \Rightarrow i < +\infty.$$

(ii) For $x < 0$, $i > \left(\dfrac{x^2-1}{2x}\right)$. Then for $x \to \infty-$, we have

$$\lim_{x \to \infty-} \dfrac{x^2-1}{2x} = \lim_{x \to \infty-} \dfrac{1-\dfrac{1}{x^2}}{\dfrac{2}{x}} \quad \text{[Putting x = -h and } h \to \infty+\text{]}$$

$$= \lim_{h \to \infty+} \dfrac{1-\dfrac{1}{h^2}}{\dfrac{-2}{h}} = -\infty \Rightarrow i > -\infty \Rightarrow -i < +\infty$$

The above two properties (i) and (ii) imply that

$$\pm i < +\infty.$$

(iii) For $x > 0$, we have $i > -\left(\dfrac{x^2-1}{2x}\right)$. Then for $x \to \infty+$ we have

$$\lim_{x \to \infty+} -\left(\dfrac{x^2-1}{2x}\right) = \lim_{x \to \infty+} \left(\dfrac{1-x^2}{2x}\right) = \lim_{x \to \infty+} \dfrac{\dfrac{1}{x^2}-1}{\dfrac{2}{x}} = -\infty \Rightarrow i > -\infty.$$

(iv) For $x < 0$, we have $i < -\left(\dfrac{x^2-1}{2x}\right)$. Then for $x \to \infty-$ we have

$$\lim_{x \to \infty-} -\left(\frac{x^2-1}{2x}\right) = \lim_{x \to \infty-} -\left(\frac{1-x^2}{2x}\right) = \lim_{x \to \infty-} \frac{\frac{1}{x^2}-1}{\frac{2}{x}} \quad [\text{putting } x = -h \text{ and } h \to \infty+]$$

$$= \lim_{h \to \infty+} \frac{\frac{1}{h^2}-1}{\frac{-2}{h}} = \infty \Rightarrow i < \infty+ \Rightarrow -i > -\infty$$

From the above two properties (iii) and (iv) we get

$$\pm i > -\infty.$$

Thus we conclude that imaginary unit iota '$\pm i$' is greater than -∞ and is less than ∞. In other words we have $\pm i < +\infty$ and $\pm i > -\infty$. These results also strengthen the concept of imaginary origin '$\pm i$' as is discussed earlier.

From above discussion we finally conclude that the imaginary unit 'i' and '-i' lie beyond the real number line]∞, -∞[and they cannot be equal to either a finite real number or infinity.

Exceptions

Although we have proved some paradoxes, we have their exceptions as follows:

Case-I: If $i > 0$, multiplying both sides by i, we get -1 > 0.

Case-II: If $i < 0$, multiplying both sides by i, we get -1 > 0.

Case-III: If $-i > 0$, multiplying both sides by - i, we get -1 > 0.

Case-IV: If $-i < 0$, multiplying both sides by i, we get -1 > 0.

Therefore real number zero and imaginary unit iota cannot be compared as far as law of trichotomy is concerned, because in every case, it gives absurd result. It doesn't mean that the present study is absurd, but is the indication that real numbers and imaginary numbers are two different number systems and their comparison leads absurd results.

Open Problems

Can do we answer the following arithmetical basic operations between imaginary unit and infinities, i. e.

$$i + \infty = \ldots\ldots?$$

$$i - \infty = \ldots\ldots?$$

$$-i + \infty = \ldots\ldots?$$

$$-i - \infty = \ldots\ldots?$$

$$(\pm i)(\pm \infty) = \ldots\ldots?$$

$$\pm \frac{i}{\infty} = \ldots\ldots?$$

Conclusions

From the above discussion we finally get the following paradoxes of imaginary unit:

(i). $i = \sqrt{-1}$ and $i = -\sqrt{-1}$.

(ii). $i < \frac{x^2 - 1}{2x}$ and $i > \frac{-(x^2 - 1)}{2x}$, for real $x > 0$.

(iii) $i > \frac{x^2 - 1}{2x}$ and $i < \frac{-(x^2 - 1)}{2x}$, for real $x < 0$.

(iv) $i > -x$ and $i < x$ for real $x > 0$.

(v) $i < -x$ and $i > x$ for real $x > 0$.

(vi) $\pm i < x$ and $\pm i > -x$ for real x > 0.

(vii) $i > 0$ and $i < 0$.

(viii) $-i > 0$ and $-i < 0$.

(ix) $\pm i > 0$ and $\pm i < 0$ for real x > 0.

(x) $i < -\infty$ and $i < +\infty$.

(xi) $i > +\infty$ and $i > -\infty$.

(xii) $\pm i > +\infty$ and $\pm i < +\infty$.

(xiii) $\pm i < -\infty$ and $\pm i > -\infty$.

References

Ahlfors L.V., Complex Analysis, McGraw Hill Book Comp., International Ed., 1979

Brown J.W. & Churchill R.V., Complex Variables, McGraw Hill Comp., Int. Ed., 2003

Copson E.T., Theory of functions of Complex Variables, Oxford Univ. Press, India, 1992

Daves Short Course on Complex Numbers

Gullberg J., Mathematics: From the birth of numbers, W. W. Norton & Company, NewYork, 87-88, 1997

Mabkhout S. A., New philosophical view merging infinity to the imaginary number, https://www.researchgate.net/publication/299804268, 2016

Mathematics Network, University of Toronto, UK

The Math Forum @ Drexel

Yadav D. K., An analysis of the imaginary unit i and its position on the imaginary number line, International Journal of Mathematical Sciences and Engineering Applications, India, 2(1), 203-209, 2008

Yadav D. K., Introduction of a Circular Number Line, International Journal of Mathematical Sciences and Engineering Applications, India, 3(2), 257-266, 2009

YOUR KNOWLEDGE HAS VALUE

- We will publish your bachelor's and master's thesis, essays and papers

- Your own eBook and book - sold worldwide in all relevant shops

- Earn money with each sale

Upload your text at www.GRIN.com and publish for free